For Helen Vendler
with best wishes,
Elizabeth Bartlett

AROUND THE CLOCK

ELIZABETH BARTLETT

St. Andrews Press
Laurinburg, N.C.

ISBN 0-932662-76-5
Library of Congress Catalog No.

First edition

Printed in the U.S.A.

Many of these poems have appeared in literary journals, including the St. Andrews Review, Crosscurrents, Antigonish Review, Pierian Spring, Candelabra, Westwords, Chandrabhaga, National Forum, Arc, Images, Stone Country, Liberal & Fine Arts Review, Crazy Quilt, Convergence, Manhattan Poetry Review, Late Knocking, Galley Sail Review, San Fernando Poetry, etc.

Over the years, the commendations in regard to her publications amount to a literary time-clock in the life of the author. To mention but 12 whose influence mark her dial, the hands point to

Richard Wilbur
William Stafford
Richard Eberhart
Jonathan Williams
Kenneth Rexroth
Robert M. Hutchins
Allen Tate
Conrad Aiken
Mark Van Doren
Louis Untermeyer
Marianne Moore
Wallace Stevens

It was their praise and support that made her books possible, her writing fellowships, her development as a poet. In acknowledgement of her indebtedness and gratitude, this book is dedicated to them.

The 12-tone poem is a new form, created by Elizabeth Bartlett. It was inspired by Arnold Schoenberg's musical system. The poem consists of 12 lines, divided into couplets. Each couplet contains 12 syllables, using the natural cadence of speech. The accented sounds of the words are considered tones. Only 12 tones are used throughout the poem, repeated various times. As a result, the poem achieves a rare harmony that is purely lyrical, enriching its imagery and meaning.

Other Books by the Author

Poems of Yes and No
Behold This Dreamer
Poetry Concerto
It Takes Practice Not to Die
Threads
Twelve-tone Poems
Selected Poems
The House of Sleep
In Search of Identity
Dialogue of Dust
A Zodiac of Poems
Address in Time
Memory Is No Stranger
The Gemini Poems
Candles

Literary Olympians, an international anthology

Contents

I.

II.

III.

I.

A CASE OF POSSESSION

I found it as a child,
a house that was all mine

Where I could think and be
whatever I believed.

Half of me stayed outside,
on guard, aware of spies

The inner self went free
to wander as it pleased.

Leaving the day behind,
I came upon the night

And there I dreamed of things
past all imagining.

NIGHT EYES

As I look through the darkened windows
of my sleep

My eyes drop below the horizon,
disappear.

It is time to leave, to follow the night
on its course

Across roofs, highways, sea lanes, skyways,
rain and snow

Pausing here, peering there, recording
its signals

Until, sun-caught, my eyes race the earth
to home base.

DREAM SEARCH

I had to go on and on,
the search was not done.

Winding corridors,
walls leading from door to door.

Mice, lions, sheep, chickens, frogs,
unassorted odds.

Nothing suited - quite -
despite the resemblances.

I heard voices, laughter, groans,
sounds foreign to mine.

Mirrors, symbols, signs...
twice, I almost found myself.

SEE YOU TOMORROW

How the bedtime refrain still echoes
through the house:

"Good night, sweet dreams, see you tomorrow!"
Was it wish

Or something more substantial for child
to sleep on

Like a pillow filled throughout the night
with promise?

Which was kept and shall be kept in years
yet to come

When all the yesterdays that made me,
wake at dawn.

A READING OF PALMS

Map in hand,
I studied the surfaces and depths

Of the land
assigned to me for exploration.

Flesh engraved,
the contours clearly showed the main routes

Time had paved
for me to follow by sun and stars.

Whether eyes
misread the signs or feet betrayed me

All the skies
my palms enclosed led far out to sea.

RECALL

Flying effortlessly
I escape gravity

And seaborne, breathe through gills
to swim past coral isles

Where I emerge on shores
that climb up ancient roads.

Always, my origins
enact some past within

Recalling elements
of former existence.

Save two, that I renounce:
bloodrust fire, fleshtorn ground.

DOWN TO THE PRESENT

I do not know
why our most distant memories

Loom the largest,
as though they started from a peak

Then descended
toward the present range after range.

Now on clear days
I see the past across my mind

Like mountain waves
receding higher and farther

The distances
rising backwards over the years.

REMIND MY CALENDAR

My memories
are always older than I am.

Parents, teachers,
the adult world in which I grew

Remain the same:
they never overtake my age.

Only children
remind my calendar of years

That I shall be
an old, unchanging memory

While they stay young
forever - remembering me.

RETURN TO THE PAST

Walking in the city of my mind
I meet friends

Who say they do not remember me
or my name.

I return to the streets I once knew
but feel lost.

Library and school have disappeared
with summer.

The trees have moved away to make room
for highways.

And I wake on the shores of my heart
cold and wet.

TROPIC ARTIST

When he gave me a painting of hammocks,
I knew:

The dreamer tells the truth, the self awake
does not.

For years I raged against the images
he drew:

How they stared, gloomy shrouds, whenever I
forgot.

To rest, be still - I swore that was a way
of death

Yet find more lives in sleep than I have years
ahead.

THE CHANCE

I dare not hold the mirror
to my images

For fear they may see
my neutrino anti-self.

Should they disappear
and I become a zero

Think what aborted talents
all of us might lose!

Better a blind course
on the undetermined chance

That I and they may harvest
some positive charge.

COMPLETELY PERSONAL

Now that time intervals
in billionths of seconds

Are exactly measured
and precisely controlled

I must find a system
designed to measure mine.

Now that all substances
are processed, analyzed

In decimals so small
they require exponents

I need a formula
to help me know myself.

GENETIC MISTAKE

An interruption
in my chain of DNA

Some latent virus
in the fibres of my nerves

Or how else account
for a black sheep in the fold?

An X in my genes,
a misreading of my code

From a cosmic ray
and a sudden shift of stars

Or how else account
for such alien ways as mine?

BELL JAR

Companion of my breath,
dear *anima mea*

Here rest, inside this glass,
invisible as air.

You are my private smile
and ring a lovely tone

That no one else can hear.
It echoes within me.

May you create your world
out of musical notes

And so content my heart
in rarest solitude.

HEAVEN

It was a lake enclosed by
an 8' bookcase

With time from now on
for reading and reflection.

An angel stood alongside
in friendly welcome.

"All the world's literature
is here," she explained.

I was enthralled, and wondered
if my poems, too...

Wordless, I stared at the words
rewarding my life.

STILL LIFE

Seduced by hand and eye,
a moment stole from time

One sunset sky and tree,
unchanged within a frame.

On my wall, the painting,
through my window, the scene.

One, captured as image,
the other, nature-free.

The tree, blown by the wind,
feels life flow through its limbs.

As rain descends dusk-grey,
I hear it breathe, drink, sway.

SKYCAVES

From this skycave
I look out and see holes of light

Red and white stars
that open passageways through space.

Against the dark
and far into the night they gleam

Their distant fires
sending signals across the years

From their skycaves
on other island galaxies.

And I wonder,
where do they lead, these caves, to what?

THE CROSSROADS

I followed God
until he stopped at a crossroads.

On the one side
was a cliff high above the sea.

On the other,
a dense woods baited with summer.

Future unknown,
I asked which way was I to go?

He pointed left,
where the sun bowed low in the west.

And God, I asked?
To the right, he said and vanished.

IN THE HANDS OF THE KEEPER

I wanted to lift
the poor and ignorant soul

To feed and clothe it,
to give it eyes and ears.

I led it away
from hunger, cold and terror

Helping it to climb,
to trust my choice of freedom.

But when I let go,
the peak opened with wide jaws

For the slip, the fall —
and I grabbed the soul, and ran.

THE UNKNOWN ONES

I am not one —
Among the many I am part.

I do not know
how many or who, where they are

Each has a name,
a face, an age — none of them mine.

Yet are we all
cells of the selfsame root in time.

No closer ties
bind us to those we call our own.

For we are one,
living each other's lives, unknown.

RIVALS

Between eros and thanatos
I am held fast.

The weight of one
is balanced against the other

By a mechanism so fine
I dare not miss

A single heartbeat or a thought
on either side.

As warmth fears cold, and light fears dark,
so my desires...

An opposition which will end
with my last breath.

MONTAGE

How clear the shadow of the earth
across the moon:

Part black, part white,
yet also marked with mud, blood, gold —

A portrait view
stark as a montage of mankind.

Against a patch of blue it glows
for all to see —

Our destinies here intertwined
to form a whole.

We are all things —
through us the universe goes forth.

THE PLUMED SERPENT

Across the cracked land
a dry river climbs upward.

From its perch on icy peaks
the sky lifts talons.

Lightning strikes. Thunder echoes.
Comes the rain for corn.

Quetzalcoatl —
a god both plumed and serpented.

I wake.
The rustle under my bed vanishes.

Two blue-winged curtains
blow through the open window.

VOICES

I spoke to God and Devil,
waiting for reply.

I whispered into both ears,
having seen two sides.

I told of men and women,
of youth and of age

Of joy, love, goodness, beauty,
and their counterface.

Each question and emotion
met with silent dread.

One listened to the living
and one, to the dead.

COMPLETION

Butterfly wings, a pair of lungs,
a bivalve shell.

I see the M and W
traced on my palms.

Maple keys, antlers, feathers, ferns,
the tails of fish —

The one design repeats itself
in endless halves.

Mountain to valley, spring to fall,
high tide to low,

We are each other's counterpart,
together, whole.

THE WHITE HERON

Each year, at winter's start,
the white heron returns.

He watches from the marsh
as night descends on earth.

Beneath his wings, the sun
is hidden from the shore

Spared the dying season
in a world left forlorn.

Yet through the darkened reeds,
time stirs an old desire

As spring and fall together meet
in bird and fire.

LOVE SONG

Sea gray, sky blue,
I soar into your eyes and drown.

Fin deep, wing high,
You are the wind, the wave in me

An urgent tide,
a flaming cloud against the night.

I sleep, I dream,
The stars set sail for your island

To hear the bird
that sings in the nest of your hair.

I float to shore
in the fluted shell of your hands.

PERSPECTIVES

Our dreamscape is a Mil Cumbres
across the years.

Peak after peak they rise like crests
above a sea

In which we plunge, swim, dive and drown
beneath each wave.

Yet breath returns and eyes grow clear
from time to time

As all stand still, becalmed, at rest,
and we can see

There, where we were. Here, where we are.
How far. Which way.

ENTER MEMORY

Memory is no stranger
in the house of sleep.

It comes as a visitor
for a reunion.

If a private occasion,
with the family

Or else with those forgotten
who have long been gone.

The waiting house is ready
for us to gather.

Together or separately
our memories meet.

SAFE JOURNEY

The twelve hours of the night
are paths between the stars.

Whichever one you take
leads to the central house

If you speak the password
to those who guard the gates.

You must not look at them
or touch them on the way

Else you'll be left outside
stopped by their triple bark.

For the rest, safe journey
and sweet dreams until dawn.

VISION

It looked like a mountain
with garden terraces

A holiday setting
and dazzling in sunshine

Where one could be at ease
to stroll and meet old friends

Exploring all the paths
unhurried by the years

Feeling the light within
increase with heightened joy

while going up and on
from terrace to terrace.

BETWEEN SHIFTS

When sleep creeps inside the sleeve
of forgetfulness

And snuggles down with a sigh
to the day's eclipse

My mind empties its pockets
of the shucks and shells

That wrap life in a garment
filled with trivia

And moves with a quiet grace
into night's shadows

Where the soft breath of candles
is the only sound.

II.

WAIT A SECOND, PLEASE

Our earth is losing
half of one second a year.

One second equals
over 9 billion cycles

As radio waves
from an oscillator prove.

We orbit the sun
by ephemeris seconds.

An atomic clock
will gain or lose one second

In an interval
of 300,000 years.

TO THE WHALES, CONDORS, WOLVES, ET AL

On Christmas trees made of wire
with silver fringes

Glass animals hang from threads
tied to the branches.

See how they twirl and glisten,
cloned by the million!

We gather around and sing,
"Joy to birds and beasts

"From men of peace and good will,
cheers, hallelujah!

"Now the endangered species
will not grow extinct."

WAVES

From the rising tide of wheels
mornings and evenings

To rolling crests of hunger
towards borders and shores

From overflows of children
with no place to go

To the flux of endless wars
whose dead never win

From the billowing of whales
who code ancient seas

To the radio light from stars
no longer sending.

A DIVIDED VIEW

A two page ad
shows both sides of the Atlantic.

Here, four children
who are protegés of science

Smile in safety,
their lives protected New World style.

There, an arrow,
drawn in red across Old World map

Points to danger,
where unseen faces hide behind

The fear of war,
all-out, global, nuclear war.

WATCHMAN, WHAT OF THE NIGHT?

1

As you eye the city's streets
for sly, swift shadows

As you scan air and sea lanes
for storm clouds and reefs

As you hunt from your tower
for signs of distress

Say, watchman, what of the night?
Say, what of the night?

Is that smoke against the sky?
What is it we smell?

Where are those sounds coming from?
Why do our hands sweat?

2

The thousand hours move slowly,
lights blink on and off

While we sit and talk
with our backs to the window

While we try to sleep
on the rough floor of our hearts.

Speak, watchman, what of the night?
Speak, what of the night?

Who are those figures
running along mountain trails?

Why are the children crying,
huddled together?

3

We do not know why
we are afraid of the night.

Who can understand
an angry wind or mad sea?

When rocks fall, we make
a cave of protecting arms.

When there is no light,
we see by maps in our minds.

Yet are we afraid.
There are canyons in the heart

And rivers of thought
that lead, watchman, who knows where?

4

Like murky negatives
searching for hidden proofs

Radar scouts the forest,
echoes pinpoint the mist.

Then guns — ripping the ground.
Rockets — splitting the air.

Till there is nothing left
but ended yesterdays.

What else is there to lose?
All that has been, we were.

Here were our homes, our lives.
Here, where we worked and dreamed.

5

A school with its roof torn
stands deaf and dumb and blind.

A cafe is open
but has nothing to sell.

A wounded dog wanders
through empty, burnt-out shops.

If there are unpaid debts,
tell us, who will pay them?

Telephone lines are down,
buses have stopped running.

Who will need them, the dead?
They no longer need them.

6

Time's cross-kneed pendulum
swings from noon to midnight

A helpless Christ, who asks
how long is forever.

Through the wreckage of glass
to the huge bomb craters

Past twisted steel girders
of bridges and highways

The darkness groans aloud:
watchman, what of the night?

And then more emptiness
fills the mouth with silence.

THE MARINER

The first one
arranged grass blades and sticks for his chart

To show winds, tides and sailing routes
across seas

And used a coconut for sextant
as his guide.

The same one
now delves into the abyss of space

With mirrors mounted on multiple
telescopes

And navigates the starred cosmos to
find God's shore.

FORGOTTEN HEROES

They were gone for days,
the hunters and fishermen

Challenging the beasts
who claimed the land and the sea.

We sang their praises
throughout the spring and summer

Our hutment ready
to give them welcoming feasts.

How their deeds warmed us
winter nights! How bright the blaze!

Now we are fearful
and cold. We miss those old men.

GOLF COURSE ON OLD INDIAN BURIAL GROUND

Did they cut down the mountain
to make a green lake?

We no longer hear bird calls
or animal cries.

We do not see smoke signals
that welcome or warn.

There is a constant thunder
without rain all day.

Then more light, more stars at night
than we ever saw.

It is hard for us to rest,
harder still to sleep.

RESPITE

The hours roll by
like a coconut on the surf

As my window
frames a canvas of tropic skies.

Such memories
as time digests to feed my mind

I turn like shells;
they are limed with careful secrets.

My favorites
are those buffered by wind and waves

To a polish.
I have a growing collection.

PRELUDE

Across the gray land
a light flush begins to spread.

The earth is menstruating.
Elm, ash, mesquite sigh

As fields tremble, yield
to the drumbeats of the rain.

Its mad music thrusts
deep inside with crack and thaw

Forcing seed to root,
to assure generation.

The sun's breath turns warm,
softening the winters's womb.

LIBERATED

Compare
worm that I was in primordial life

With the woman today, conscious of place
and time

Then note-
the direction has moved on up, not down

Up from algae, with a determined mind,
my own

Shaped by millenia and meant to climb
more heights

Leaving the fossil worms to prove how far,
how wise.

AN OLD ROAD

It was a wornout trail
that stretched the eye for miles

With whitecapped thunderclouds
streaking toward the mountains.

We brushed past the snowflakes
disguising fields of corn

While tall cosmos flowers
offered us the summer.

When we came to a stream,
it was silent and dry

But an old stone basin
in pity shared its drink.

ADIOS

When planes did away
with the soft wings of angels

And you needed a ticket
to get to heaven

When God was shot to his death
by a computer

and the devils were released
from hell back to earth

We invented a time clock
and bought a burro

Then built our own shrine
and learned to make tortillas.

HINDSIGHT

Clouds, shadows, fog...
such the intrusions I question

The obscure events that sharpen
the fear and loss

Remembering what was before,
without a doubt.

Can things remain the same or not,
the past return?

Or will moods change and seem unreal
when the scene clears?

Sky, weather, view...
I seek a vanished world in me.

INCOMPARABLE

More than taste to the tongue
though it serve me mulled wine

More than scent to the breath
though it perfume my nights

More than touch to the hand
though it feel love's heartbeat

More than sound to the ear
though all music is speech —

The speed of eye's own light
on earth, to outer space

And self's creative thought
that forms my universe.

BRAIN ROCK

Scattered over the sand,
the mica chips of thought.

Buried in mountain walls,
the gleam of massed ideas.

Pebbled by coursing streams,
the solvent math of time.

Beneath the pressured earth,
the pure crystalline view.

Such knowing as the mind
assays from outer world

Turns object bone to stone,
the atom's human weight.

SUBJECT AND OBJECT

One says it matters,
the other says it doesn't.

With the stars as numerous
as granules of sand

How can people make
a particle of difference?

I have enough negatives
in my universe

To know it takes positives
to fill up the holes.

And then what has no matter
does matter later.

ABOVE ALL

The physical... the animal...
most certainly.

We have all the basic symptoms,
why deny them?

The tyranny, the chemistry
of bones and glands

Assert our nature's evidence
by force of will.

Yet how explain beliefs based on
angels and saints

Unless the concepts they produce
are also proof?

AMBITION

How hungrily it feeds,
the greedy worm, desire

A leech that can not wait
to banquet in the grave

But fattens on its host
through living days and nights.

Ah, when its maw is filled,
boring through skin and bone

Will slave go free at last
to claim some rich reward?

Or years be recompensed
with a crown for the corpse.

SUNDAY PAINTER

Twenty blue dragonflies
and one redwinged hopper

While the lake slants the sun
toward a tall clump of weeds.

My dog plays at leap-frog,
each splash a foot away

While the sky sails three kites
above the mountain's rim.

I sketch and move about
with restless hand and eye

While a green orchid vine
leans on the wind and swings.

PETAL-LIFT

A carnation so beautiful,
it seemed unreal

Pale, with a glow almost too pure,
too fresh and smooth

Its slim neck erect, its spine tall
as a proud girl

Breathing an intimate perfume
best sensed close by.

How to keep it that way a while,
to make it last...

Floral cosmetic surgery —
surely that could?

UNDENIABLE

I think, therefore I am
served Decartes' doubt as proof

While I release my poems
as prayers rather than truths.

If these should be denied,
then so must I be, too.

For who can be believed
and not be heard or seen?

Can time outside of mind
exist unless perceived?

Only God knows for sure —
and He, He thought Adam.

BY OUR WORDS

To say that God is dead
is to say we are dead

For whether His image
is ours or we are His

We are surely mourners
at our own funeral

And the womb of heaven
is changed into a tomb.

Living ghosts, we wander
across a stillborn world

Waiting for a rebirth
to prove that we are real.

PETITION

Demon or angel,
I seek your powers and charms:

In an insane world
to keep my sense of reason

In the cave of dreams
to find peace without terror

In my spleen and heart
to overcome frustration

In times of fortune
to bless or curse its outcome.

So might I go, bare
as when I was born, but whole.

EXCELSIOR

From nightingales
to buses, trucks and motor cars.

From laurel wreaths
to Oscars and Emmy awards.

From Socrates
to computers and recorders.

From oracles
to gold, oil and Wall St. prices.

From nymphs and gods
to the Pentagon and Kremlin.

O world be praised
for all the progress we have made.

CHOICES

Between the mass and minority,
take your stand;

Private passions or public problems,
please decide.

Protest, criticism and revolt —
that's one way.

Silence, patience and resignation —
that's another.

To Kalon said the Greeks and poets,
looking up.

Panem et circenses said the gods,
looking down.

III.

PENTIMENTO

She wanted to obscure
her past and to change it

To conceal the mistakes
and unwanted background

Beyond recognition,
with artistic finesse.

In time the pigments dulled
and like a facade, peeled.

It was a disturbing
revelation, to find

The former print still there,
still visible, alive.

THE MERMAIDS

Caught in a sea of nets,
the years become tangled.

They reach through kelp and moss,
they struggle to be free

Pleading for light and breath
and to swim with the tide.

Mermaids of memory,
they sun themselves on rocks

Enjoy sky and flowers,
sense new taste, sound and touch

Until... pale, stiff and dry,
they slip back to the depths.

REUNION

This was the night we planned
our summer reunion

Here, in a private cove
beneath the salt-streaked cliffs

While an El Greco moon
stared down through leaded clouds.

We smoothed the sand of stones
and poured a libation

Praying the tide to wait,
to let us drink and dance

Before the flood of years
make us forget, forget.

HEART'S HOUSE

A house unloved, unlived in —
how cold, how lonely!

No welcoming caresses
to warm its bare walls.

No echoing cries to rouse
its windows, floors, roof.

Just ghosts through colorless days,
searching signs of life

Just memories from old graves,
pitying the nights.

Heart, my house of empty rooms,
where has my love gone?

HARD LUCK

I write words
that mortar together like small stones

Not to throw
at the world, to raise up welts and cries

But grounded
for walking on, for holding me up.

(To endure
becomes a matter of survival).

I pick words
to sound them out, see if they will burn

As I do
on this rough road that has been my life.

NIGHT THOUGHTS

When the days proceed on ifs
that may or may not.

If desert, I choke on thorns;
if sea, I shall drown.

I turn and toss in the wind;
my feet twist on stones.

I am tangled in a fog
between walls of sleep.

The hair on my neck grows damp;
my nails grow sharper

While I tear the hours of night
into small dark words.

DISTURBING THE ORDER

An angry butterfly,
a flower gnawed by grief —

Is it any wonder
they feel sick and depressed?

A bird lost in its flock,
a cub that can not howl —

Is it any wonder
they hate and blame themselves?

Let's admit disorder
is failure turned to fear.

A case of neurosis,
psychic identity.

THE FIRST LAW

It's the rejection of self
that is hard to take.

It's the worst kind of failure,
beyond retrieval.

Days turn cold and gray with age
under sunless skies.

The moon's forever hidden
in a starless night.

No shadows or reflections
to protest the lie.

Self, who can say what you are,
if here you do not?

THAT RARE ELEMENT

"You need lithium to keep calm,"
says the doctor.

"Too many ups and downs upset
nature's balance.

"A benign equanimity
is best for health."

I think of mountains and valleys
changed to plateaus

Of rivers, brooks and waterfalls
born motionless

Of how it is, in self defense,
that rocks can melt.

VOICELESS

I have silenced my Philomel —
unhappy child!

No words speak her agonized love's
forgotten joys.

In her mute eyes some once bright spark
swims in darkness.

No song now breaks the dawn's membrane,
oh desert days!

A piece of glass. A red feather.
No more, no more.

I veil my throat and seal my ear
to harsh incries.

CHIMERA

I embrace the young ghost
of a warm memory

And find the kisses cold,
the arms sharp and brittle.

Oh let go! warns my brain.
And forget! cries my heart.

How well they know, they know
the goat of my fancy

Who rules a pantheon
of immortal lovers.

I have no more flowers.
The last candle burns low.

"AND OH, THE DIFFERENCE TO ME!"

I thought love spoke in one language
needing no words

I felt love was a way of life
one and no more

I was sure love had its own world
one clear and true

But some hearts beat louder, faster
like captured birds

And some feet have a longer stride
meant to explore

While some eyes wake from sleep and dreams
alone, alone.

THE LAST STRAW

By the time I noticed,
it was a barren field

The harvest long gathered
and the reapers long gone.

Only the wind to tell
how much was chaff or husks

Empty as memories
devoid of promises.

Still, there might be something
that had been overlooked

But no one heard the cry
when the earth broke in two.

THE GUINEA FOWL'S TEARS

She makes sure no one sees them,
despite the dark rings

That enlarge her eyes
with a sad intensity.

She knows how hard a substance
is reflecting glass

How unsubstantial
the image made on water.

Better to mourn in private,
wear unspotted black

With white beads for prayer,
feathers for flights from the heart.

STONE

Deep in the nuclear core
beyond listening

There beats the lithic force
of some inert fury.

Wrapped inside its pyrene dark
and mutely screaming

It cries for someone's help:
to be let out and live

To feel again the passion
of being seen, heard.

Within its fossil heart
a love lies atrophied.

J'ACCUSE

How angry you were back there
looking towards the front

Knowing you could not be seen
or heard - and who cared!

Well, as Dostoyevsky said,
sometimes God pities

A man who at last despairs
of his lot and life.

But now you are successful
and as blind and deaf

You do not like reminders.
Am I God, you ask?

LURES

Butterflies and moths —
both find accommodation

In the days and nights
that fill our twin existence.

So too ice and fire
that alternate in our veins

Between impotence
and tremulous desiring.

Confused by the light,
we are drawn to what awaits:

Substance or shadow,
daydream or carbon nightmare.

HERE AND THERE

Here, the village —
each life encircled by a wall

Of family, friends and neighbors,
a solid fort

Unless a crack, then a fracture
and the sky falls.

There, the city —
each self but a small, single brick

Among hundreds of tall buildings
vying for space

While crowded dreams rise like bubbles
ready to burst.

PRECONCEPTIONS

The cat, hurled down from the roof
in a game of boys

The dog, left behind one night
when we moved away

Made life a yowl of terror,
of wept–out, blind eyes.

The fish, transferred from a bowl
to a bright new tank

The bird, found hurt on the road
and placed in a cage

Made death a floating belly,
a cold, feathered stick.

RESILIENCE

They skate on boards that roll
on straight or crooked paths.

They waltz, they pirouette
on tall trim legs of chance

Indifferent to bones
that move and bend with ease.

Gravity becomes them
as they trip, lurch or sprawl

Then bounce up on their feet
to commence a new dance.

We lose grace with the years,
we forget how to fall.

CHILDREN

We were children of light,
then children of darkness.

Now children of shadows,
we endure and survive

As boneless images
who risk both sticks and stones.

Penumbras make it clear:
the flawed drives out the pure

Whether apples, money
or demon memories.

We bend, flex and sidestep,
perfect imitations.

NEW MUSIC

They come rocking to the festivals
snapping feet

Or sit around bonfires toasting songs
quietly.

They hold hands to form a no-nuke fence
with their lives

Or link a continent by marching
for their rights.

They bring a new music to the world
with their heat

Gathering the fragments that promise
them some peace.

SILENT FILM

How do you tell them
not to make the old mistakes?

Can you teach the heart to break?
How do you reach them

Through the walls of words that lie
in mouth, ears and eyes?

Must our shadows dance
on the dark screen of their minds

Where the feet move blind?
Must some dream unseal by chance

The long reel of anguished nights?
O silence, be light!

ISLANDS OF ICE

Slowly the cold and the rain
formed a glacial mass.

Blinded by snow storms,
no one noticed when it moved.

Cut off from home base,
the family broke apart

Like icebergs at sea;
one went mad, one froze, one drowned.

Another built an igloo
out of stocks and bonds.

There are no monuments
for burials at sea.

QUESTIONS WITHOUT END

When will the words be opened
and the book unsealed?

Not till the time of the end
of empires and beasts.

When will the dream be written
according to men?

Not till the signs and visions
have become as one.

How shall we learn to know them
as true evidence?

Not till all human senses
merge with light again.

SNOW

1

This whiteness conceals
the center of our being.

Exposed to the cosmic draft,
our nature profaned

A willful alteration
upsets our balance.

The whiteness does not shield us
from the sun's ions,

Earth darkens beneath the threat
of endless winter

While our center shifts
slowly, mechanically.

2

We venture dangerously
into the future

Exchanging heart space
for an outer dimension

Daring the whiteness
to make us as pure again.

A cold logic perverts us
as weightless beings

Who grow bodiless
in regions of gravity.

Around us revolves
an invisible abyss.

3

In a destitute time
we swing uncertainly

From fact to fantasy
like puppets in a void

Enticed by the sayers
who would save us with words

Or lured by the doers
dancing us to exits.

Anything seems better
than freezing in darkness.

Anything seems better
than nothing's nothingness.

4

Oh for a rising wind
to sweep the snow away!

Oh for a breath of love
to dissolve ice and frost!

For we can not roam the forests
as beasts again

Nor recall the mythical gods
to rule again.

Yet sure as earth's axis
keeps turning it towards light

So our human axis
must lead us through this night.

THERE ARE LIMITS TO KNOWING

I can not understand
existence without love.

I hold the prejudice:
nothingness is unreal

Like the death of matter
in a negative world.

The air in which birds fly
makes me aware I breathe

Like them, for my own sake,
but conscious what we share.

Fish do not know water —
they live and die in it.

JOURNEY'S END

A lost ship
I moor in the shelter of your arms

And hold close
with frayed rope, torn sail and leaning mast.

Dependent
on wind, chance and my own woman's strength

I have voyaged
far and long from point of origin

To this port,
the safe, blessed haven of no return.

Now keep me
content to dream, at peace on the shore.